Reliability Centered Maintenance

Overview and Best Practices

John Ciulla MSME.
Owner: TRP Services LLC
ASNT Level III Vibration Analyst
Six Sigma Black Belt

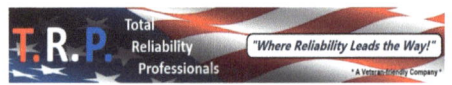

Table of Contents

Table of Figures

Table of Acronyms

API	Asset Productivity Improvement
CAPA	Corrective Action Preventive Action
CBM	Condition Based Maintenance
CDT	Condition Diagnostic Testing
CM	Corrective Maintenance
CMMS	Computerized Maintenance Management System
EC	Equipment Criticality
EFS	Equipment Functional Assistance
FA	Failure Analysis
FMEA	Failure Modes Effects Analysis
HAZOPS	Hazards and Operability Studies
HHP	High Hazard Process
ICD	Instrument Control Drawing
IDLH	Immanent Detriment to Life and Health
ISO	International Organization for Standardization
JSA	Job Safety Analysis
MEL	Master Equipment List
MI	Mechanical Integrity
MIP	Maintenance Improvement Plan
MTBF	Mean Time Between Failure
O&M	Operations and Maintenance
OD	Operational Description
OEE	Overall Equipment Effectiveness
OEM	Original Equipment Manufacturer
P&ID	Process and Instrumentation Diagram
PdM	Predictive Maintenance
PM	Preventive Maintenance
PMA	Preventive Mitigation Analysis
PSM	Process Safety Management
QA	Quality Assurance
RAG	Risk Assessment Grid
RCFA	Root Cause Failure Analysis
RE	Reliability Engineer
RCM	Reliability Centered Maintenance
SFR	System Functional Ranking
TBA	Time-Based Activity
TPM	Total Productive Maintenance
WO	Work Order

1. INTRODUCTION

Reliability Centered Maintenance (RCM) is both a Maintenance philosophy and a methodology.

- The "reliability-centered" philosophy focuses all maintenance activities and planning on preserving the design reliability of any physical asset.

- The RCM methodology is a process to determine specific maintenance requirements necessary for a desired reliability level in the current operating environment.

Properly executed RCM will help optimize the use of maintenance resources.

RCM seeks to minimize to improve reliability throughout the asset life-cycle by using techniques such as improved design, Preventive maintenance, and Predictive Maintenance tools such as condition monitoring. RCM applies maintenance strategies based on consequence and cost of failures.

This approach allocates maintenance resources in areas that have the greatest leverage on reliability including plant safety and quality.

1.1   RCM Overview

The RCM process is used by Reliability Engineers, maintenance and operators to answer the following seven questions:

1. What are the functions and associated desired standards of performance of the asset in its present operating context/ (Functions)

2. In what ways can the asset fail to fulfill its functions? (Functional Failures)

3. What causes each functional failure? (Failure Modes)

4. What happens when each failure occurs? (Failure Effects)

5. In what way does each failure matter? (Failure Consequences)

6. What should be done to predict or prevent each failure (Proactive Tasks and Task Intervals)

7. What should be done if a suitable proactive task cannot be found? (Default Actions).

✎ The core steps in the RCM process are as follows:

- Perform Asset Criticality Analysis
- Prioritize equipment for RCM process
- Execute the RCM process

✎ For each piece of equipment or system to be analyzed:

- Compile failure experience
- Complete a failure modes and effects analysis (FMEA)
- Conduct prevention/mitigation analysis (PMA)
- Create Master Asset Management Plans
- Identify critical spares

1.2   Methodology

There are two overall approaches to RCM

Classical/Rigorous RCM

a. *Benefits*: Classical or rigorous RCM provides the most knowledge and data concerning system functions, failure modes, and maintenance actions addressing functional failures of any of the RCM approaches. Rigorous RCM analysis is the method first proposed and documented by Nowlan and Heap and later modified by John Moubray, Anthony M. Smith, and others. This method should produce the most complete documentation of all the methods addressed here.

b. *Concerns*: Classical or rigorous RCM historically has been based primarily on the FMEA with little, if any, analysis of historical performance data. In addition, rigorous RCM analysis is extremely labor intensive and often postpones the implementation of obvious condition monitoring tasks.

c. *Applications*: This approach should be limited to the following three situations:

- The consequences of failure result in catastrophic risk in terms of environment, health, or safety, and/or complete economic failure of the business unit.
- The resultant reliability and associated maintenance cost is still unacceptable after performing and implementing a streamlined type FMEA.

- The system/equipment is new to the organization and insufficient corporate maintenance and operational knowledge exists on function and functional failures.

Streamlined/Abbreviated/Intuitive/ RCM

a. *Benefits*: The streamlined approach identifies and implements obvious improvements based on "sufficient" analysis, RE experience and operator/maintenance inputs. In addition, it culls or eliminates low value maintenance tasks based on historical data and Maintenance and Operations personnel input. The intent is to complete enough rigorous initial analysis in order to realize early-wins that help offset the cost of the FMEA and proactive activities such as condition monitoring. In summary, this approach follows the so-called "action leaning" improvement model: i.e. Analyze, Apply, Realize gains, Learn and repeat the cycle to the point of Diminishing returns.

b. *Concerns*: Reliance on historical records and personnel knowledge can introduce errors into the process that may lead to missing hidden failures where a low probability of occurrence exists. In addition, the intuitive process requires that at least one individual has a thorough understanding of the various condition monitoring technologies.

c. *Applications*: This approach should be utilized when:
  - The function of the system/equipment is well understood.
  - Functional failure of the system/equipment will not result in loss of life or catastrophic impact on the environment or business unit.
  - For these reasons, the streamlined or intuitive approach has been recommended for government applications such as DOS (Department of State) facilities, NASA, and NAVFAC (Naval Facilities Engineering Command) operations. In addition, a streamlined approach has been successfully used in both discrete and continuous manufacturing facilities.

Either approach to RCM defined above can be applied in one of two ways, depending upon local circumstances. These two methods are referred to as (1) Equipment-Level RCM, and (2) Systems-Level RCM. These applications are further defined below.

1.2.1 Equipment-Level RCM
In this method, individual pieces of critical equipment are selected for analysis. Resources are focused on one piece of equipment at a time.

Examples of effective uses of this approach include the following:

ᗌ Analysis involving many pieces of identical or very similar critical equipment

ᗌ Analysis of systems that consist of a very few pieces of equipment

ᗌ Analysis of a relatively few pieces of equipment that are **known** to be preventing the process from reaching desired rates, reliability, costs, etc.

This approach can require a considerable amount of time and resources for large numbers of critical equipment. Also when using this individual equipment approach the impact of individual equipment failures and remedies on the entire asset "system" may not be obvious during the initial analysis.

1.2.2 System-Level RCM

This approach begins with the selection of systems within a plant that will be subjected to the RCM process. These systems are then prioritized, so that the most critical ones are analyzed first. The approach in this case is to focus on a group of equipment that is functionally interrelated. A FMEA is completed for the entire system rather than individual FMEAs for each piece of equipment. Prevention / mitigation analysis is then performed for the critical equipment within the system.

Advantages of this approach include the following:

ᗌ More effective use of time and resources

ᗌ Greater participation across different functional areas (Production, instrumentation and electrical, process engineering, etc.)

ᗌ The effects of a malfunctioning piece of equipment on the system are determined.

When using the systems method, in certain circumstances, one iteration of the analysis may not be sufficient. An additional iteration of one or more subsystems may be required to fully result in the identification of maintenance tasks that will fully meet the process objectives. In these cases, a second cycle may be required.

1.3 End Product of RCM

A *Reliability-Centered Maintenance (RCM)* program will result in the optimum mix of reactive, time- or interval-based, condition-based, and proactive maintenance practices. These principal maintenance strategies, rather than being applied independently, are integrated to take advantage of their respective strengths in order to maximize facility and equipment reliability while minimizing life-cycle costs.

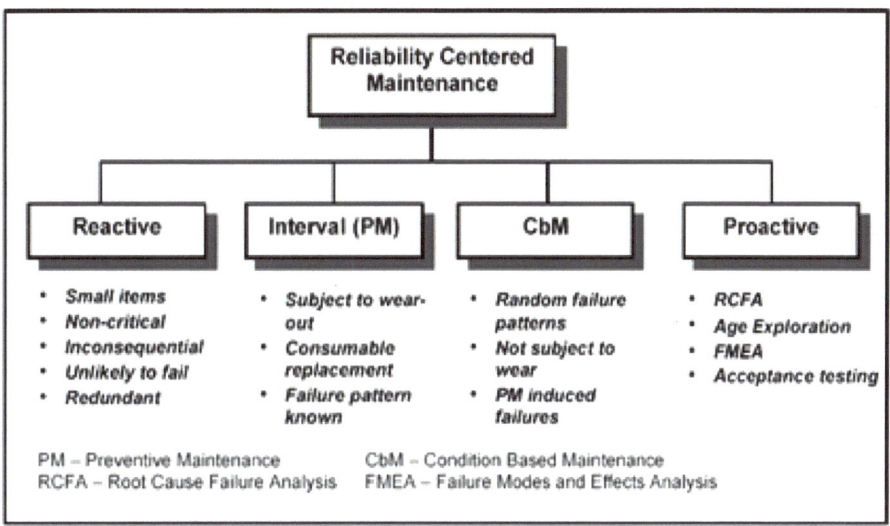

Components of an RCM Program ⇧

RCM includes a blend of reactive, time-based, condition-based, and proactive tasks. In addition, a user should understand system boundaries and facility envelopes, system/equipment functions, functional failures, and failure modes, all of which are critical components of the RCM program.

The RCM end product is an O & M Master Plan for a physical asset or asset system. For installed equipment this master plan may include some previous maintenance activities (others may no longer be necessary and new maintenance activities. The obsolete tasks are removed from the CMMS or made inactive.

1.4 Implementation

The RCM process is not for everyone and implementation requires significant time and resources. Therefore, in order to get the greatest benefit (ROI), careful prioritization of systems and/or equipment is necessary. Operations and maintenance leadership must participate in this prioritization to assure that Reliability resources are well focused on supporting key objectives. This prioritization must be repeated at regular intervals to account for changing circumstances and operational needs.

RCM will normally be used only on "business critical" systems and equipment. While systems-level RCM is adequate for most needs, in special circumstances such as PSM (Process Safety Management) or Quality-critical applications, equipment-level RCM may be necessary. Typically less than 20% of the failure modes are responsible for 80% of the problems. Look for the bottlenecks, the recurring failures, and follow the money.

Reliability Engineers are responsible for developing and implementing the RCM implementation strategy for their area. They are supported, as necessary, by process engineers, operations personnel, planners, craftsmen and others who have knowledge of historical performance and quality requirements.

The local Reliability Engineers are also responsible for determining which RCM approach described above is appropriate for any given application.

1.5 Change Management

Implementing an RCM Program involves significant business process and personal behavioral change. The most common situation is one of changing from a reactive maintenance environment to a proactive maintenance approach.

Successful navigation of this transition requires four things – without exception.

☞ **Executive Sponsor** – someone with organizational "clout" to help create the vision, marshal resources, and neutralize interference when resistance to change happens. This is the person with "corporate connections." *Axiom:* "when trying to change anything in an organization always have two levels of support above the person who needs to change."

☞ **Participative Plant Resources** – local Maintenance, Operations, Engineering, IT, Finance (they count the gains) and other key support groups must be invited to engage and then be deeply involved in all work. This creates ownership. *Axiom*: "human beings in any organization will not destroy that which they build."

☞ **Process** – building an RCM program requires: 1) an end point, 2) a series of steps to get there and 3) logical starting point. Education is essential. Typically, this program must be explained many times – *Axiom:* "it takes 10 units of communication to get one unit of understanding when changing anything".

☞ **Metrics** – the customer (Operations & Finance) must be able to prove that RCM makes the operation run better. This is best done by tracking improvement in a few KPIs. Solution:
- find the Management KPIs (i.e. bonuses get paid on them)
- Link all RCM work to those KPIs.
- Establish a baseline (to create an unambiguous starting point)
- Track changes – the numbers must be "bullet proof" (Hint: Finance loves "bullet proof" numbers and will go to great lengths to help create them – if asked).
- Report Changes – weekly to inform plant and corporate – monthly summaries for Management

2. RCM Prioritization Process

This is the process used to determine the priority order in which equipment will be analyzed using systems-level or equipment-level RCM:

a. Define and list all systems.
b. Establish clear system boundaries.
c. Perform preliminary prioritization of equipment.

These steps are discussed below.

2.1 Define and list all systems

For these purposes, a system is defined as "a group of equipment functionally related and designed to produce a certain output or service." A system function is a service, or output, of the system to a plant area.

Process flow diagrams (PFDs), process and instrument drawings (P&IDs), instrumentation control drawings (ICDs) can be used to help define the function of a system. For large facilities, it is advisable to start with a process flow diagram since a system can have a number of associated P&IDs and/or ICDs.

Technical packages usually have operational descriptions (ODs) and associated drawings.

The existing Master Equipment List (MEL) in the CMMS and ISO documentation also serve as guides for determining systems.

Typical examples of systems are:

- Equipment with a standby or backup (pump and standby pump)
- Group of identical or similar equipment in parallel (multiple plastic injection machines)
- A piece of equipment that requires other equipment to provide a service or output (cooling tower with associated pumps)
- Established conventions used on the plant floor (mixing, blending, extrusion, etc.)

A process flow diagram may be all that is necessary to define the systems. It is advisable however, to gather all available P&ID's and ICD's and other documentation at this point. The documentation will be used to further define systems and in the development of the master equipment list.

2.2   Establish clear system boundaries

Documents such as the system operational descriptions, process flow diagrams, process and instrumentation drawings, technical packages and vendor manuals can be used to identify the boundaries of the system. "Marking up" drawings is a good way to record system boundaries. Include all the components that are required for the proper function of the system. However, be aware that the functional boundaries of the system may frequently cross into other related systems. It is important to establish and maintain clear, consistent boundary conventions in order to assure that no equipment is inadvertently missed.

The following provides examples of boundary conventions:

- Active equipment, including tanks and instruments, is included with the system they serve, regardless of any other designation.
- Heat exchangers are considered in the system including the service (load) side.
- Motors are included with their driven equipment.
- Utilities and groups of equipment that serve multiple systems or production lines should be considered systems by themselves.
- Field devices for distributed control systems (DCS) and programmable controllers (PLCs) are a part of the equipment on which they are located.

After boundaries are established for each system, list on the PFD the applicable P&IDs and ICDs that will be used for developing the MEL. For large systems, it may be necessary to build a spreadsheet or database to list the P&IDs and ICDs by title and drawing number.

It is also advisable to list all available information by document type, title, number, and for controlled documents, source and location. This will make a ready reference for planners, MRO and others.

When possible, obtain electronic copies or list the location of controlled, read-only information that can be linked to the CMMS.

2.3 Perform critical ranking of systems and equipment

Begin this process by first assigning a **failure occurrence code (FOC)** to each piece of equipment. The following is an example set of **FOC** criteria. However, each area may choose to define their criteria. The following table contains examples of quantifying values often used with failure occurrences.

Category	Level Description for Frequency
Frequent	Greater than 1 per week
Probable	1 to 4 occurrences per month
Occasional	1 to 11 occurrences per year
Remote	1 to 4 occurrences in 5 years
Improbable	Less than 1 occurrence in 10 years

Figure 1: Typical Variables Used to Describe Failure Occurrence

The following table contains a score for each of the respective FOCC's.

Failure Occurrence	Score
R = Remote	1
I = Improbable	3
O = Occasional	5
P = Probable	7
F = Frequent	9

Each piece of equipment within the functionally significant systems must be assigned a **failure criticality (FC)**. This defines the severity of the failure to the system. The **FC** criteria are defined below:

Failure Criticality (FC)	Score
This failure is least severe and results in an annoyance, but equipment still functions	3
This failure can be an annoyance leading to functional failure	5
This failure is most severe and results in a functional failure	7

Figure 2: Failure Criticality (FC) Criteria

The **equipment functional significance (EFS)** can now be calculated. The **EFS** is the product of the **failure occurrences (FOCC)** and the **failure criticality (FC)**:

$$EFS = FOCC \times FC$$

Equipment whose failure will not have significant impact on unit operation is normally considered "run-to-failure". Maintenance on run-to-failure equipment is limited to inspection, periodic lubrication, default failure detection tasks, and repair or replacement when the failure is detected.

The **adjusted EFS score** takes the component EFS rating and determines how well the customer's current equipment task list addresses the component failure and ranks it. A component's **adjusted EFS score** can then be used to prioritize work. The higher the score, the more attention it may deserve. The end goal is a prioritized list of equipment, reflecting the following:

- Safety and health issues **(IDLH is always highest priority regardless of any other perceived priorities)**
- Product quality impact
- Regulatory and certification issues (OSHA, EPA, MI, ISO/QS9000)
- "Green lines" or PSM/HHP
- Downtime (or reduced Up-time)
- Maintenance costs
- Production -- actual versus design

Adjusted EFS	Scale
Component is not listed in the current task list	EFS Multiply by 3
Component is listed but does not address specific failure	EFS Multiply by 2
Component is listed and addresses specific failure	EFS Multiply by 1

Because this process depends heavily upon local facility needs, the prioritization criteria and methodology must be established locally.

If equipment histories are available in the CMMS or in operator logs, Pareto analysis of the failure data can be performed to determine which equipment has high repetitive failure rates with attendant high maintenance costs – so called "bad actors". If failure records are not available simply ask the Maintenance first line supervisors and technicians.

In making judgments, use real data when it is available as opposed to anecdotal information. Some equipment failures can be categorized very dramatically and perceived as problematic and of high priority, when in actuality, they may have a small impact on KPIs compared to other chronic problems.

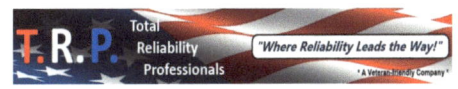

3. EQUIPMENT-LEVEL RCM ANALYSIS

The steps in the equipment-level RCM analysis are as follows:

3.1 Perform Pre-FMEA research

The RCM Failure Modes and Effects Analysis (FMEA) is a method for determining the source cause of an unsatisfactory condition and its effects on the system. Its purpose is three fold:

a. To change an asset strategy – To include condition monitoring tasks & master lubrication route – Add FTEs to the lubrication route as needed – Also include equipment changeover process documentation

b. To update task lists for each of the identified "bad actors" or critical equipment

c. Create or update the list of spare parts needed for the task.
   - Include spare parts information into the material section of the task list
   - Include photo's, drawings, key OEM contact information, etc. in the task list

Data for the preliminary analysis should include a review of the equipment history – both oral & written:
   - Relevant quality, safety, cost, production and regulatory information
   - Equipment failure investigations
   - MIR analysis
   - Lubrication analysis
   - Existing task lists and CDT routes
   - OEM equipment evaluations
   - CAD drawings and manufacturer service manuals
   - Previous FMEAs / RAGs
   - CMMS BOMs

The Reliability Engineer (RE) is responsible for this data collection. It is important to gather as many details as possible on past equipment failures. Special note should be made of the historic failures that were viewed as having a high enough probability of reoccurrence and/or severe enough consequences that preventive maintenance practices were instituted as a result of them. Attempt to capture details on what failed, how it failed, how often it fails and the effect of each failure.

3.2 Develop Preliminary FMEA

An example of an FMEA sheet is shown below in figure 3

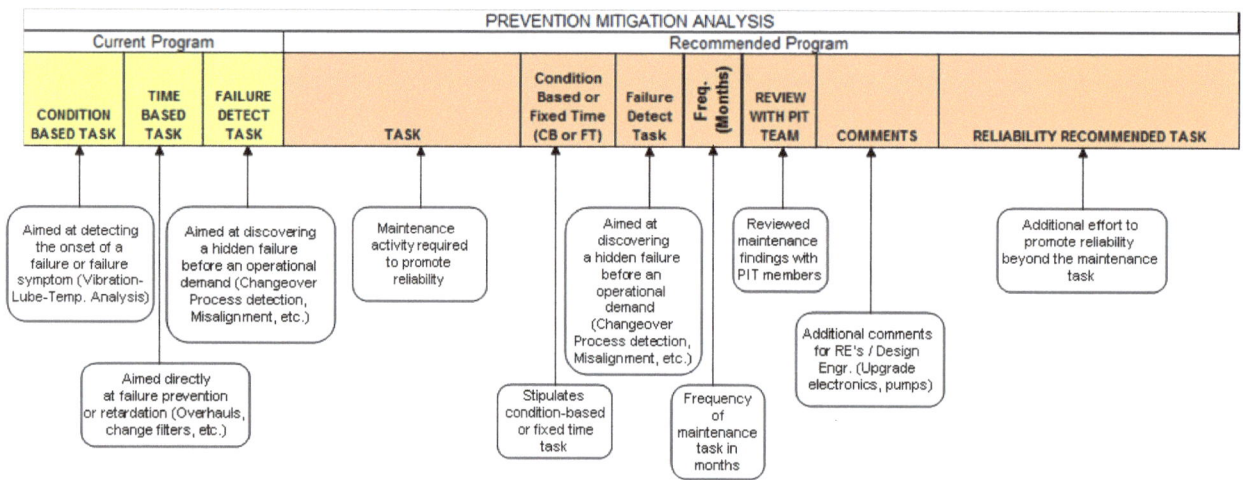

Figure 3: Sample FMEA

3.3 Perform FMEA (RAG)

It is suggested that a review of the maintenance requirements of any asset should be done by small teams which should include at least one person from the maintenance function and one from the operations function. The seniority of the group members is less important than the fact that they should have a thorough knowledge of the asset under review.

In view of this, the name for these small groups will be called **PIT** (Process Improvement Team) **Crews**. The make-up of an ideal PIT Crew is shown below:

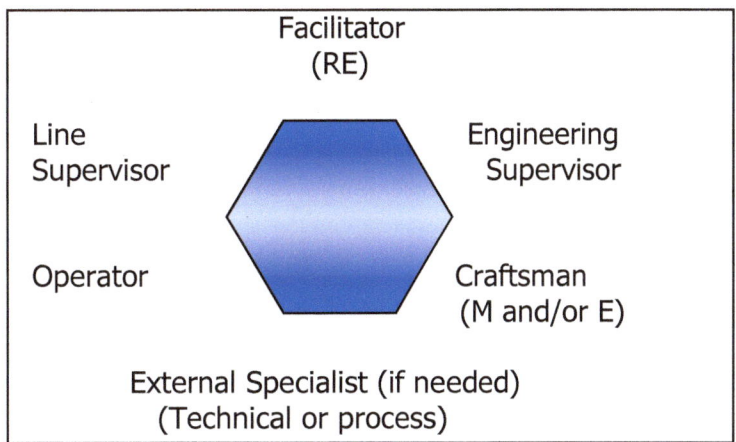

Facilitators

The facilitator role is to ensure that:
- ❑ The system boundaries are clearly defined, that no important items are overlooked and that the results of the RAG analysis are properly recorded.
- ❑ RCM is correctly understood and applied by the team members. (80/20 rule)
- ❑ The team reaches consensus in a brisk and orderly fashion, while retaining the enthusiasm and commitment of individual members.
- ❑ The analysis progresses reasonably quickly and finishes on time.

3.4 The outcome of an RCM –RAG Analysis

As stated earlier, the RAG analysis should result in three tangible outcomes:

a. To change an asset strategy – To include condition monitoring tasks & master lubrication route. – Add FTEs to the lubrication route as needed. – Also include equipment changeover process documentation.

b. To update task lists for each of the identified "bad actors" or critical equipment.

c. Create or update the list of spare parts needed for the task.
  - o Include spare parts information into the material section of the task list.
  - o Include photos, drawings, key OEM contact information, etc. in the task list.

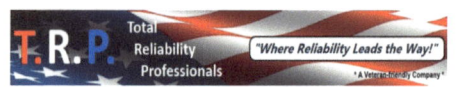

Two less tangible outcomes are that participants in the PIT meetings learn a great deal about how the asset works, and also tend to function better as teams.

 a. While doing this exercise, remember that for all but the most complex equipment, there are relatively few failure modes.

 b. Determine the functionality of each piece of equipment in the system. Those that do not directly impact the safety of the process or quality or quantity of the product or service of the system will be set aside for later application of either "run-to-failure" or "failure-finding" maintenance modes.

 c. Create an FMEA (RAG) sheet for each of the remaining pieces of equipment within the system.

 d. Enter the known historic failures in the Failure Modes column of the FMEA form.

 e. Next, in the occurrence column, enter the occurrence code that corresponds to the observed failure frequency (see example System Prioritization Matrix).

 f. In the Failure Cause column list the root cause of each failure (an RCFA may be required to complete this step).

 g. Finally, enter the effects of the failure on the system, including secondary damage, in the Failure Effects column.

3.5 Perform PMA

Prevention/mitigation analysis first identifies the maintenance tasks that are applicable to a specific component failure. These tasks are then evaluated to determine which one or combination of several will most effectively prevent or mitigate the component failure. Once a task is selected, calculations must be performed to determine if the selected task will be cost-effective. The tasks generally fall into three categories:

 a. Condition-based tasks

 b. Time-based tasks

 c. Failure detection tasks

3.6   Create Maintenance Tasks

The RE writes tasks with input from members of the PIT Crew. Questions frequently asked at this point are, "Why write tasks prior to approval? Isn't this extra work if the proposed Master Plans are not accepted?" First, generating the total package helps insure that the important details have been thought through. Second, management needs the complete picture in order to approve the plan. Finally, much of the work necessary to generate the cost estimates can be directly applied to the master plan development.

In fact, the entire master plan can be entered into CMMS and left with the PM schedule inactivated until approved. Once entered, the master plan can then be easily printed out and attached to other documentation to form a complete package for consideration.

3.7   Include new tasks in the CMMS

Immediately after the review has been completed for each asset, senior managers with overall responsibility for the equipment must satisfy themselves that decisions made by the team are sensible and defensible.

After each review is approved, the recommendations are implemented by incorporating maintenance schedules into maintenance planning and control systems (revise task list in CMMS) and by incorporating operating procedure changes into SOP for the asset (Training).

3.8 Incorporate New tasks into Planned Work

Generally assignment of preventive maintenance tasks can be summarized as shown in the guidelines below:

Type of Task	Performed By
Preventive maintenance tasks requiring use of CDT (i.e., vibration analysis, thermography,meggering,lubrication analysis, predictive ultrasonic testing, etc.), *where only light hand tools are required.*	Technician/CDT (if advanced tools are required)
Visual inspection, checking lube oil levels, and meter readings, *where only light tools are required.*	Operators
Preventive maintenance tasks requiring use of CDT where *substantial rigging or use of heavy tools are required*	Maintenance and Technician/CDT
Other preventive maintenance tasks	Maintenance

Figure 4: General Assignments of Preventive Maintenance Tasks

4. SYSTEM-LEVEL RCM ANALYSIS

4.1 Select system to be analyzed

- o Reliability Engineers (REs) define and list systems – using RCM tree or MEL as guidelines.
- o Establish clear system boundaries. Attempt to set system boundaries such that maximum 10 to 15 pieces of equipment constitute a system.

4.2 Conduct initial system prioritization

The RE prioritizes systems based on the following:
- o Safety and Environmental Issues
- o Maintenance Burden
- o Production Relevance
- o Regulatory Issues

4.3 Perform Pre-FMEA research

a. Collect all existing failure and maintenance history, CMMS focus reports, existing PMs and CDT routes, prints, job procedures, safety, cost, production or regulatory information.

b. Detail the equipment failures known to have previously happened.

4.4 Conduct preliminary FMEA

a. While doing this exercise, remember that for all but the most complex equipment, there are relatively few failure modes.

b. Determine the functionality of each piece of equipment in the system. Those that do not directly impact the safety of the process or quality or quantity of the product or service of the system will be set aside for later application of either "run-to-failure" or "failure-finding" maintenance modes.

c. Create an FMEA (RAG) sheet for each of the remaining pieces of equipment within the system.

d. Enter the known historic failures in the Failure Modes column of the FMEA (RAG) form.

e. Next, in the occurrence column, enter the occurrence code that corresponds to the observed failure frequency (see example System Prioritization Matrix).

f. In the Failure Cause column list the root cause of each failure. This should be based on previous failure investigation or general knowledge.

g. Finally, enter the effects of the failure on the system, including secondary damage, in the Failure Effects column.

4.5 Perform PMA with input from PIT Members

a. Use the System Prioritization Matrix as a guide to determine the level of maintenance resources appropriate to eliminate or reduce the risk associated with the failure types identified in the FMEA(RAG) analysis.

b. Consider how the current maintenance practices address these risks. Evaluate those practices for their contribution to reliability in terms of adequacy, effectiveness, and cost.

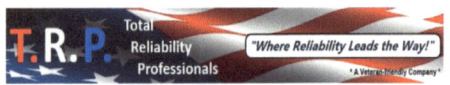

    c. Decide if adjusting inspection intervals or details can further optimize these practices. Consider whether certain maintenance practices should be discarded altogether or replaced with something more appropriate. Look for conditions where it would be advantageous to apply condition based maintenance rather than time based.

    d. Lastly, consider "run-to-failure" or "failure-finding" maintenance modes for equipment that does not directly impact the system's safety, quality or output.

4.6 Map out task package

    a. Group preventive maintenance activities by type or location where feasible.

    b. Determine which activities can be performed by which craft.

    c. Determine roll-up sequences or "packages", i.e. those specific tasks which must be performed at the same time (when a piece of equipment is made available).

    d. Organize routes to minimize time an energy expended by maintainers and operators.

4.7 Incorporate New tasks into Planned Work

Generally speaking, Operators should perform preventive maintenance work that requires no more than incidental use of small tools. Examples of these types of work include; visual walk-down, cover and guard removal for belt or chain drive inspection, oil sample collection, general use of CDT tools and meter reading.

Preventive maintenance work that requires more extensive use of tools should be performed by the appropriate craft. An example of this type of work would be inspection or other testing where large equipment housings must be opened.

5.   Master Equipment List – Hierarchy Guidelines

The MEL provides a structured consistent method to document each asset element within a facility. It is the foundation for any asset management system. The completed MEL is loaded and resides in the CMMS. It is used to track asset history including failures and repairs. The MEL can be useful when performing Criticality Analysis.

It is used to produce a list of critical equipment. FMEA's are then developed for this equipment.

The completed MEL is loaded into the CMMS. After the Hierarchy is entered into the CMMS, consistency will help assure work orders are issued and history is stored at the correct level.

5.1   Standardization

The examples and conventions listed here should be regarded as the standard approach to developing MEL hierarchies throughout the plant.

Standardization of the tree offers the following advantages:
- o RCM experience has shown that certain elements present in the plant should be placed at the equipment level. Standardization ensures that these types of elements be assigned to the appropriate level. This will facilitate the plant-wide information sharing of FMEA's for common pieces of equipment. Since the function of common equipment may vary, reevaluation of the system "effects" portion of the FMEA in order to assure that their specific pieces of equipment are properly analyzed and documented.

- o RCM analysis makes use of standard criteria definitions for each area to assign criticality ranking at the system and equipment level. Standardization of the hierarchy assures that these criticalities are consistently applied throughout the plant.

  ○ When work orders are completed by the craftsmen the results / repairs made are entered. This is done by entering a code in the appropriate CMMS field chosen from a list of components involved, conditions found and actions taken (CCA codes). Standardizing CCA codes at the equipment level permits queries of available historical data on a particular class of equipment repaired throughout all facilities. For instance, one can determine how many bearing failures of a particular type of pump have been experienced.

5.2 Levels of the MEL Hierarchy

Level	Description	Example
Area	A plant facility identified by product.	Extraction, Filaments, Utilities, etc. (Depends on Plant Type)
System	A group of equipment functionally related to produce a certain output or services.	Mixing, extrusion, material handling and conveying systems, hydraulics, etc.
Equipment	A set of components that provides a stand-alone function.	Pump assembly, extruder, mixer, dryer, compressor, column, etc.
Component	A sub-assembly that is integral to the equipment but has no stand-alone capabilities.	Speed reducer, electric motor, hydraulic cylinder, controller, valve, instruments, etc.

Figure 5: Levels of the MEL Hierarchy

5.2 Examples of Plant Conventions

5.2.1 System-Level Conventions
  ○ Utilities that support multiple pieces of like equipment are always at the system level.
  ○ A group of interrelated equipment working together to provide a service is considered a system.

5.2.2 Equipment-Level Conventions

- o Individual units such as pumps, fans, cooling towers, tanks, mixers, air compressors are always classified as equipment.
- o Electric motors, no matter what size, should be identified at the component or part level and should always be functionally associated with their driven piece of equipment.
- o Motor Control Centers (MCC's) are pieces of equipment. The equipment includes the following components: cabinet, distribution bus, breakers, safety switches, motor starters, supply power cables, instrumentation, low-voltage distribution panels.
- o A piece of equipment can include the following components (other components not listed can also be present):

Hoist			
Prime Mover	**Driven Component**	**Speed Reduction**	**Braking**
Motor, including local disconnect and starter (not a part of MCC).	Hoist Drum	Gear Reducer	Electric Brake, Load Brake

Figure 6: Typical Equipment Components

- o Control valves and field devices are components of the associated equipment.

5.3 RCM Boundary Conventions

Item	Boundary Convention
Motors	Include power cables, control circuit, local disconnect and starter, pushbutton station. Motors are always classified at component level.
Hydraulic Valves	Solenoid is included with the hydraulic valve. The valve is included with the hydraulically operated equipment prime mover that it controls.
Instrument Control Loops	Include primary measuring devices, input signal conditioning equipment, controllers, computer generated set points, output signal conditioning equipment, and final control elements.

Relays	Relays are physically included in equipment controlling the coil or relay.
Heat Exchangers	Heat exchangers are components included with the product system demanding the need of the exchanger.
Hydraulic Components	Valve stands, pilot air, accumulators, pulsation dampers are components of a hydraulic system.

Figure 7: RCM Boundary Conventions

5.4   Example of MEL Hierarchy Breakdown for a Common Plant Element

This section provides partial hierarchy breakdowns for plant elements.

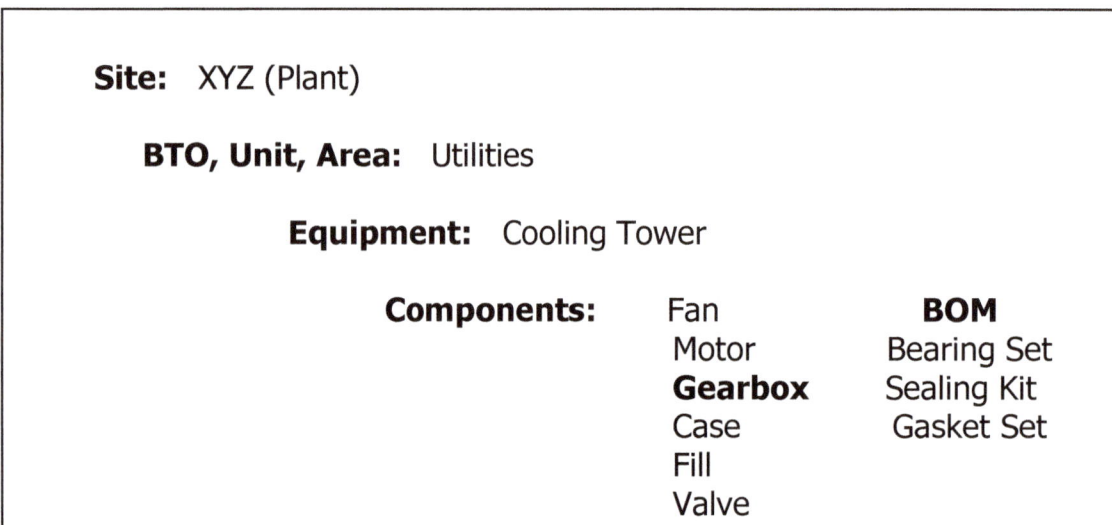

Figure 8: Partial MEL Hierarchy for Cooling Tower

5.5 Typical Breakdown for Equipment in Master Equipment List

```
0001  BAGHOUSE
      0001A  BAGHOUSE ELECTRICAL CONTROLS
      0001B  BAGHOUSE LOCAL CONTROLS
      0001C  BAGHOUSE BACK PANEL CONTROLS
      0001D  BAGHOUSE DESK PANEL CONTROLS
      0001E  BAGHOUSE SWITCHES
0002  BAGHOUSE I.D. #1 FAN UNIT
      0002A  BAGHOUSE I.D. #1 FAN ELECTRICAL
      0002B  BAGHOUSE I.D. #1 FAN MOTOR
      0002C  BAGHOUSE I.D. #1 FAN UNIT
      0002D  BAGHOUSE I.D. #1 FAN SPEED REDUCER
0003  BAGHOUSE I.D. #2 FAN UNIT
      0003A  BAGHOUSE I.D. #2 FAN ELEC CONTROLS
      0003B  BAGHOUSE I.D. #2 FAN MOTOR
      0003C  BAGHOUSE I.D. #2 FAN ASSEMBLY
      0003D  BAGHOUSE I.D. #2 FAN SPEED REDUCER
0004  BAGHOUSE MODULE #1
      0004A  BAGHOUSE MODULE #1 CLEAN AIR PLENUM
      0004B  BAGHOUSE MODULE #1 PULSE SYSTEM
      0004C  BAGHOUSE MODULE #1 DIRTY AIR PLENUM
      0004D  BAGHOUSE MODULE #1 BAG SHAKER/VIBRATOR
      0004E  BAGHOUSE MODULE #1 OUTLET PIPING
      0004F  BAGHOUSE MODULE #1 SCREW CONVEYOR
      0004G  BAGHOUSE MODULE #1 ROTARY VANE FEEDER
0005  BAGHOUSE MODULE #2
      0005A  BAGHOUSE MODULE #2 CLEAN AIR PLENUM
      0005B  BAGHOUSE MODULE #2 PULSE SYSTEM
      0005C  BAGHOUSE MODULE #2 DIRTY AIR PLENUM
      0005D  BAGHOUSE MODULE #2 BAG SHAKER/VIBRATOR
      0005E  BAGHOUSE MODULE #2 OUTLET PIPING
      0005F  BAGHOUSE MODULE #2 SCREW CONVEYOR
      0005G  BAGHOUSE MODULE #2 ROTARY VANE FEEDER
0006A  BAGHOUSE INLET DUCT
0006B  BAGHOUSE OUTLET DUCT
```

Figure 9: Standard MEL Breakdown for a Baghouse

<u>RCM Summary</u>

      Some of the major "take aways" from my RCM publication would be to always research best practices, benchmark current facilities with like processes and implement industry best practices. When you couple this process with safety you will create a world class harmonization that will give you increases in total plant efficiency and reliability!

      My 15 years of experience beginning as a Predictive Maintenance Technician through becoming a seasoned Reliability Engineer has provided me the skills and knowledge to implement Total Reliability Centered Maintenance. Bringing reactive facilities to a proactive, preventive and overall more efficient state.

      Thank you for reading my RCM initiation handbook and hopefully by implementing some or all of these techniques, you will become champions in RCM and Total Asset Management!

Regards,

John Ciulla MSME.

www.trp-services.com

"Being challenged in life is inevitable, being defeated is optional"

Memo Notes:

Memo Notes: